I0414789

Cosmic Origin of Dust and Humanity

Jin He

authorHOUSE®

AuthorHouse™
1663 Liberty Drive
Bloomington, IN 47403
www.authorhouse.com
Phone: 1-800-839-8640

First published by AuthorHouse   11/30/2009

ISBN: 978-1-4490-5108-2 (sc)

Printed in the United States of America
Bloomington, Indiana

This book is printed on acid-free paper.

Cosmic Origin of Dust and Humanity

– Origin Exploration 2 –

Jin He

Contents

List of Figures

Chapter 1

Cosmic Origin of Dust

1.1 Human Crisis on the Earth

The history of human on the Earth is at least several hundred million years. However, we do not know how human is originated. Human beings seem to be the wandering street children who have lost their parents. The most horrible thing is that human beings never realize that they are the wandering street children. We know neither our blood lineage nor the meaning of life. We do not know how to value our own or other's lives. Such blind and cruel actions as deceit, plunder, warfare, and so on, have strangled millions and generations of lives.

Today, the global world needs a reflection on itself and mankind needs to resort to its starting point. It is time for human beings to seek their origin and answer such primary questions as what human body is made of and what humanity is. Further more, the crises that the Earth's inhabitants encounter require mankind's answer to these questions!

Towards the 21st century, mankind on earth encounters unprecedented crises. The natural environment suffers the most important crisis which may reclaim mankind's fate. The crisis manifests itself with air and river pollution, ecological imbalance, as well as the global warming caused by greenhouse effect. The second crisis is the financial and social crisis caused by human's bubble economy. The resolution of these crises turns out to be mutual contradiction. In addition to these crises, the ever-changing material life has also brought the devastating effect on human body and mind.

Reviewing through human history, we can see that the crises stem from human's ignorance of their own origin. If human beings know the origin, then all kinds of crises can be prevented, the existing crises can be mitigated and even eliminated.

Very fortunately, human understanding of the nature over the last centuries is enough for human beings to understand the origin. Physics, chemistry, biology and medicine have fully proved that human bodies are composed of elementary particles. The question is, what force combines elementary particles into the biosphere?

1.2 The Power which Governs Everything: Gravitation

Scientists have fully proved that there exist only four forces among particles: electromagnetic, weak nuclear, strong nuclear, and gravitational. The nuclear forces are short-ranged while the electromagnetic force is long-ranged. Each of the three has two contradictory aspects of attracting and rebeling and, generally, has no net effect in the macro-world due to offsetting effect. By artificially destroying the offsetting effect, scientists and engineers can make scientific or commercial products based on the earth's natural structures. Atomic bombs result from artificially destroying the offsetting effect of nuclear force. Computers, telephones, TV sets and so on are the examples of artificially destroying the offsetting effect of electromagnetic force.

Gravitational force, however, has no contradictory aspects. Gravity has the only effect of attraction and, therefore, can not offset itself. Because of this, the true origin of natural structure is the gravitational force.

The origin of natural structure can not be

other forces. Modern science has fully proved that independent system of microscopic particles combined by electromagnetic force or nuclear forces inevitably moves towards chaotic state rather than orderly one. This is the principle of entropy increase, which is well known for scientists. Therefore, if there were no gravitational force then the whole universe would be simply uniform gas without structure. However, there exist in the macroscopic world such orderly structures large as galaxies and small as stars, planets, plants, animals, and even human beings. Therefore, varied kinds of macro-world structures result from the struggling of the gravitational force against the electromagnetic and nuclear forces.

Unfortunately, gravity is very very weak. For example, it is $0.0000 \cdots 00001$ times (where 40 zeros are after the decimal point) weaker than the electricity! Only the Earth, Moon, Sun and so on present gravity. There is no slightest gravity between cars or human bodies. Therefore, human beings suffer insurmountable difficulty to experimentally study gravity. Because human beings are insignificant, it is impossible to do physical experiments on such macroscopic systems as the solar system or galaxies. Even if human beings could do such experiment, they would not have sufficient time to complete it. We should know that the life of the Sun is about billion years!

However, we can use man-made telescopes

to take images of large-scale material systems such as galaxies. We can analyze the images!

1.3 Human Body Material: Dust

Human beings can understand the gravity by studying the large-scale structures in the universe. Both human bodies and planets are mainly composed of the elements which are heavier than hydrogen and helium. On the contrary, the Sun is mainly composed of the lightest elements, i.e., hydrogen and helium. Therefore, human beings could neither live on the Sun nor be originated from it. However, cosmic dust is similar to human bodies in their constituents, and is mainly composed of the heavier elements. Therefore, any planet in the universe must be derived from cosmic dust. Of course, humans beings must be derived from cosmic dust. This confirms what Bible says: man is made of earth! Therefore, if we know the origin of cosmic dust, we must know the origin of humanity!

Very fortunately, it is very easy for us to see where is dust or no dust by looking into the image of any physical system in the universe. The basic constituents in the universe are galaxies. Therefore, knowing how dust is generated in galaxies is equivalent to knowing how lives are originated. And we can even know more. We can even know what humanity is! This is what the book talks about.

Do not be fooled with color images. Some people are indulged in women's pretty looks, but they simply do not know what is color. Color is essentially the different frequencies or wavelengths of light. In fact, the shape of an object or its photo is the distribution of light arriving at your eyes from the surface of the object. That is, it is the distribution of light frequency and density varying with the surface of the object. Light of longer wavelength that appears reddish has strong penetrating ability. In other words, reddish light refuses to be absorbed by dust or gas. Elliptical galaxies are very clean, with no observation of gas and dust. Therefore, it does not matter to catch which color for you to take the images of elliptical galaxies. Images of the same elliptical galaxy of different colors are very similar and smooth. They are the good demonstration of star distribution in the galaxy. But elliptical galaxies are three-dimensional while their images are two-dimensional. The image of an elliptical galaxy is the cumulative density of stars in the observing directions.

Spiral galaxies are just the opposite. They have a large amount of gas and dust. Although their shapes are two-dimensional, they have a certain degree of thickness. Therefore, if we take images of spiral galaxies in the shorter wavelength (i.e., bluish light) then the light from the stars that are behind gas and dust are basically absorbed by the gas and dust. As a result, the image is mainly the distribution of gas and dust. Because the distribution of gas and dust is not smooth, the image looks ugly. Internet images of spiral galaxies are usually short-wavelength ones, therefore, people are daunted by the mysterious look of gas and dust (see the lower-panel of Fig. 1.1). Therefore, to get an image of spiral galaxy which is mainly stellar density distribution, we take light of longer wavelength from the galaxy, e.g., infrared image. The resulting image is reddish. Although gas and dust have charming and bright colors, they have negligible mass.

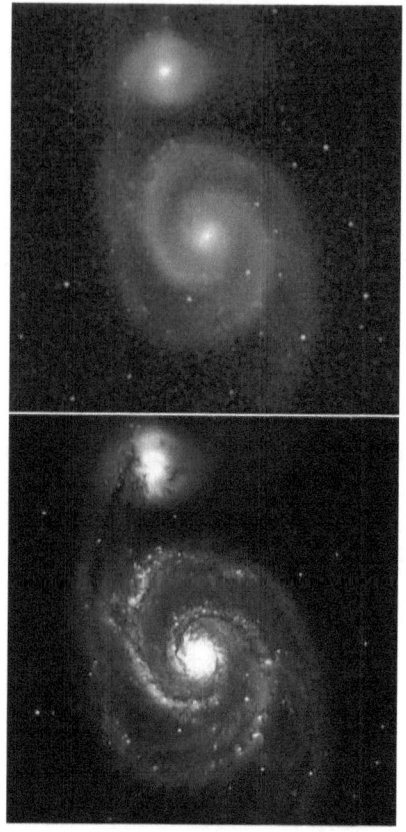

Figure 1.1: Upper panel: the infrared image of normal spiral galaxy M51 (image credit [1]). Lower panel: the blue-band image of the same galaxy (image credit [2]).

1.4   The Origin of Galaxies: Rational Gravity

The usually familiar gravity refers to the force which exerts between Earth and Moon or between Sun and Earth. These examples of gravity are the interaction between two bodies. As for the behavior of gravity exerted on many bodies, the solar system can not be the example for us to study such behavior. However, each galaxy is composed of billions of stars, which demonstrates the gravitational interaction among many bodies. Galaxy images show that each galaxy has a center. Star density at the center is the highest. From the center outward, the density is smaller and smaller and presents a regular pattern, known as galaxy structure. The principle behind the formation of galaxy structures is the demonstration of gravitational interaction in many-body systems.

Then, what is the behavior of gravitational interaction in many-body systems? Dr. He pioneered the study on galaxy structures in 2003 and the study shows that stars in any galaxy are controlled by a very simple orderly force involving many-bodies: proportion. Because solar system is just a point at the Milky Way galaxy, the proportion force reduces to Newtonian gravity between two bodies!

Proportion means that the distribution of matters in the universe is orderly. For example, there are four giants standing in array. Their heights are respectively A, B, C, D, and A, B stand in the first row from left to right, C, D in the second row from left to right. According to the view of mainstream cosmologists, the four giants can have any heights and can stand at any position. That is why current foundational scientific theories are incompetent and can not explain the origin of natural structures! They can not provide any basic principle to explain such orderly structure as human beings nor to resolve the motion of the most simple gravitational systems (such as interactional free three-bodies). However, the universe is orderly. The orderly force at the largest scales requires that the distribution of heights is in proportion. In other words, A divided by B is equal to C divided by D. This means that A divided by C is equal to B divided by D. If there are nine giants standing in array, then the ratios of heights from neighboring two rows are constant (proportion rows). Similarly, the ratios of heights from neighboring two lines are constant (proportion lines). In this way are galaxies created!

The above-said rows and lines are all straight (proportion lines). but each galaxy is a regional distribution of matters in the universe. Therefore, the proportion lines of each galaxy are curved but the rows and columns still cross at each other vertically and they form the net of orthogonal curves.

In general, a distribution of similar bodies is called the rational structure if its density varies proportionally along some particular net of orthogonal curves. In book 1 of Origin Exploration [3], rational structure is called the matriarchal structure. From now on we call it rational structure. Therefore, independent galaxies are all rational structures. The force which leads to the rational structure is called the rational force, i.e., the proportion force which is the demonstration of gravity in large-scale and many-body system. That is, the universe is rational. This is the most important discovery in human history.

1.5 Proposition: Rational Structures are at Most Bilaterally Symmetric

This is a mathematical proposition: any net of orthogonal curves is either circularly symmetric with respect to the center point, or bilaterally symmetric. I spent three years from 2002 to 2005 to look for a net of orthogonal curves whose shape has odd symmetry. That is, I wanted to find a rational structure which resembled a two-arm spiral pattern like the spiral galaxy M51. The three-year study indicates that a net of orthogonal curves is generally connected to some complex analytical function. As you might know, a complex analytical function always has two parts (real and imaginary), and leads to a net of orthogonal curves. However, complex analytical functions are very special ones which satisfy some strong conditions like Cauchy integral theorem and formula. From my experience, I do not find any complex analytical function whose graph of the real or imaginary part has odd symmetry. Therefore, I have the proposition that rational structures are at most bilaterally symmetric. Please help me prove the proposition.

Surprisingly, galaxy structures happen in the same way as indicated in the following.

1.6 Coincidence: Galaxy Patterns (except Arms) are at Most Bilaterally Symmetric

Amazingly, any component of any galaxy pattern (except the arm pattern) is either circularly symmetric, or bilaterally symmetric. And my academic papers [4-11] show that, except the arm structure, any component of any galaxy structure (such as exponential disks, galactic bars, even the whole elliptical galaxy) is a rational structure. That is, any galaxy image (ignoring the arm) can be fitted identically to rational structures.

The following will present more and more

cases of coincidences. Therefore, they are not coincidences at all. They are the cosmic truth, and Dr. He's discovery is absolutely important!

1.7 Coincidence: Dust and Irrationality

It is the observational fact that spiral galaxies are full of dust while elliptical galaxies have no dust. Why does this happen? However, people have not found its answer since galaxies were discovered more than 80 years ago. Dr. He gives the answer.

Arm structure is neither circularly symmetric with respect to the center point, nor bilaterally symmetric. Therefore, arms are not rational structures. Arm pattern tends to be oddly symmetric with respect to the center. But we can never find such "grand design" arm pattern which presents the perfectly odd symmetry. In fact, there are very different types of spiral structures. Some galaxies (like M51, Fig. 1.1) are what we call "Grand Design" spirals, meaning that they have a clearly outlined and well organized spiral structure. Other galaxies, like NGC 4414 are called "flocculent" and have much harder to trace arms. Compared with the exponential disks, galactic bars, and even the whole elliptical galaxies which demonstrate smooth and rational structures, arms may not be called a structure at all.

Arm patterns exist only in spiral galaxies and they are weak compared with the main disk structure of spiral galaxies. Therefore, the presence of arm structure is the disturbance to the rational structure. Because arm patterns exist only in spiral galaxies and only spiral galaxies present dust, Dr. He comes to the critical answer to the above question: any disturbance to rational structure must produce cosmic dust!

1.8 The Origin of Cosmic Dust: Impulsive Gravity

From now on we call the disturbance to rational structure the cosmic impulse. Therefore, the universe is originated not only rationally but also impulsively! In the case of large-scale structures (that is, galaxies), the impulse is demonstrated to be the disturbing waves, i.e., the arm patterns.

We have seen that the rational structure is the main structure while the disturbance to the main structure is always weak. That is, rational force is the main one while the impulse is weak. In the case of large-scale structures (that is, galaxies), the disturbing waves try to achieve the minimal disturbance and, as a result, they follow the proportion rows or columns of the rational structures. The impulsive disturbance reveals the rational design of the universe: proportion!

Because we live inside a galaxy (Milky Way galaxy) and we find no other force exists except gravity at our neighborhood, we conclude that Newton or Einstein formulation of gravity is a partial result of the universal "gravity" and the real gravity presents not only rational sense but also impulsive one as indicated at the large-scale galaxy structures!

1.9 Further Evidences that Dr. He's Discovery is Absolutely Right!

A: The exponential disk of any spiral galaxy is a rational structure which is circularly symmetric about the galaxy center. Mathematical calculation indicates that the proportion curves of the rational structure are equiangular spirals which are observationally the curves represented by normal spiral galaxy arms.

Figure 1.2: Upper-left panel is galaxy NGC 2983 ([12], see Evidence C). Upper-right is the infrared image of galaxy NGC 1365 ([1], see Evidence D). Lower-left is the infrared image of NGC 1300 ([1], see Evidence F). Lower-right is NGC 1365 (ultraviolet image).

B: Galaxy arms are oddly symmetric about the galaxy center. Mathematically we can not find any rational structure which is oddly symmetric about a center point. Therefore, galaxy arms are not rational structures. Observationally, they are the disturbing waves to the main rational structure (exponential disk). In order to achieve minimal disturbance, the waves follow the proportion curves of the rational structure. As a result, disturbance reveals the elegant design of the universe: proportion. Disturbing wave is a kind of active order. We call it impulsive order. Therefore, the disturbance of the active order to the rational order produces gas and dust in spiral galaxies. Gas and dust give birth to new stars and their corresponding planets. We human beings live in such a spiral galaxy. Elliptical galaxies, however, do not allow the survival of wave disturbance. Accordingly elliptical galaxies are very clean without gas and dust. The stars in elliptical galaxies are long-lived and very few new stars are nurtured.

C: Rational structures are usually circularly symmetrical about the center points. The only rational structure we can find which is not circularly symmetric, is the bilaterally symmetrical structure, namely, dual-handle structure. Observationally, only two kinds of spiral galaxies are found in the universe. One kind of spirals are the normal spiral galaxies which are composed of only disks and arms. The other kind are the barred spiral galaxies. Amazingly astronomers do observe the dual-handles in barred spiral galaxies (see upper-left panel in Figure 1.2). Dual-handle structure is also called the sub-bar of barred galaxies because a galaxy bar is usually composed of two or more sub-bar structures (see Figure 1.2 and 1.3).

D: The main structure of spiral galaxies is the exponential disk. When the dual-handle structure (i.e., sub-bar) is near the galactic center, the superposition of the dual-handles to the bright central disk presents a bar shape. This precisely explains the origin of galaxy bars. A galaxy bar is usually composed of two or more dual-handle structures. Observationally, there are barred spiral galaxies which present two nonparallel sub-bars (see upper-right panel in Figure 1.2).

E: Compared with exponential disks, bars are observationally weak structure. That is, bar structure is so weak in the outer areas of any spiral galaxy that it is ignored. Amazingly mathematical calculation of dual-handle structure shows that it is weak when compared with the disk (see Figure 1.3)!

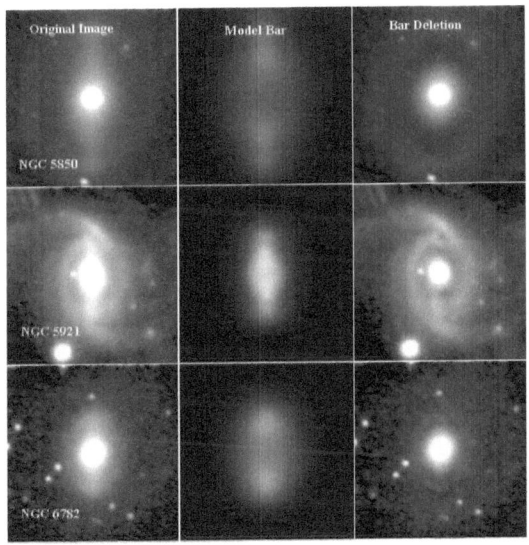

Figure 1.3: The simulation of galaxy bar with dual-handle structures. The images [13] minus our model bars respectively result in the disk and bulge images (bar deletion).

F: There are mathematically spiral-shaped proportion curves in dual-handle structure. However, they are not equiangular because they surround the central line of the dual-handles (recall that the spirals in exponential disks are equiangular and surround the center point). Two proportion curves which are oddly symmetric about the center point in dual-handle structure make approximately elliptical shape and its long axis must be parallel to the central line of the dual-handles. Surprisingly, astronomical observations show that arms of barred spiral galaxies do surround the middle lines of their bars, and they are not equiangular spirals, and the two arms make approximately elliptical shapes with the long axes being parallel to the bar middle lines (see Figure 1.2).

G: Mathematically, exponential disks have circular proportion curves. Observationally, some normal spiral galaxies do have closed arms which are circular, called rings. Mathematically, dual-handle structures have the closed proportion curves which are ellipses whose long axes must be parallel to the central lines of the dual-handles. Observationally, some barred spiral galaxies do have closed rings which are ellipses and the long axes are parallel to the galaxy bars.

H: The simulation of galaxy bar images with dual-handle structures is very well (see Figure 1.3).

I: Dr. He has proved that elliptical galaxies are completely rational structures of three-dimensional shapes [14]. The proportion curves of elliptical galaxies are the intersecting nets of orthogonal spheres, where disturbing waves are difficult to form and spread. On the other hand, spiral galaxies are two-dimensional and their proportion curves are open spirals where disturbance waves are easy to form and spread. Astronomical observations do show that arms do not exist in elliptical galaxies.

The disturbance to rational structure leads to the formation of gas and dust. New families of stars and planets are born to these gas and dust. The star-planet families are short-lived. This happens only in spiral galaxies.

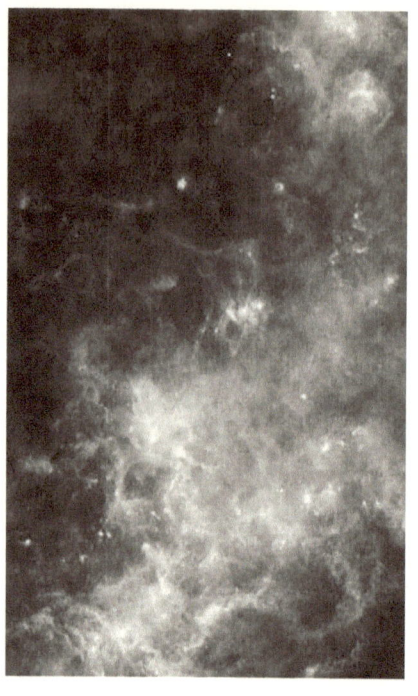

Figure 1.4: The image (credit [15]) shows that the Milky Way galaxy, a spiral galaxy, is in "a very turbulent process", constantly creating new generations of stars.

1.10 Rationality and Impulse: the Origin of Life

From the above we see that elliptical galaxies have only rationality and got no impulse. Therefore, they present no living phenomenon. There is no supernova explosion, no dust in elliptical galaxies. They are composed of the same kind of long-lived stars, and each star is almost void of life-supporting elements, i.e., the elements which are heavier than hydrogen or helium. Spiral galaxies on the contrary, are subject to impulsive disturbance, and present dramatic activities of cyclical life. Take a look at the images of spiral galaxies: supernova explosion, flying dust, colorful and rich production of heavy elements which maintain life!

The above is the large-scale phenomenon taking place in the universe. Its rationality is the absolute proportion and its impulse is the grandiose disturbance waves. Such rationality and impulse create the Primary Product of the universe: cosmic dust!

Chapter 2

Cosmic Origin of Humanity

2.1 The Primary, Secondary and Tertiary Products

Cosmic dust is the Primary Product of the universe. If there were no dust, there would be no planet. If there were no planet, there would be no such high-level life as human beings. Mankind is the Secondary Product of the universe. Further, all known products on Earth such as languages, cultures, money, construction, commodities, and social structures are created by human beings, which are the Tertiary Product of the universe.

Very few people realize that life is the phenomenon of impulse to rationality. Furthermore, many people do not regard themselves as organic materials. Similarly, very few people have realized that man-made products (Tertiary Product) such as computers, telephones, atomic bombs, etc., are usually inorganic. They are the products of pure human reasoning and have no impulse themselves. Instead they are the tools with which the Secondary Product (human beings) express their own emotion. Tertiary Product generally has no impulsive emotion of their own.

2.2 Children: the Pure Product of the Natural Power

Human beings are the Secondary Product of the universe who was bred by the Primary Product (i.e., cosmic dust) under some special natural conditions. Of course, the Secondary Product (human beings) shares the common living nature as the Primary Product (cosmic dust): rationality and impulse. A human fetus that has not yet been born has already have its rationality and impulse. Many adults contempt babies and consider they are naive. In fact, they are naive with respect to the capacity of knowledge based on the Tertiary Product created by adults themselves. Human beings are the natural organic computers of the universe and babies are the brand new computers who have not stored much information on the Tertiary Product. Adults are the old computers who have stored much of the man-made information. Some items of the Tertiary Product are human's spiritual viruses. Many adults are crushed by these viruses. For example, money is one example of the Tertiary Product and it is sometimes viral to human spirit.

Acquisition of the cosmic truth is equivalent to the installation of anti-virus software to human spirit. Cosmic truth is the truth about human origin. Human crises are resulted from humans' ignorance of cosmic truth. If human beings know their own origin, then all kinds of crises can be prevented, the existing crises can be mitigated and even eliminated.

2.3 Living in Three dimensional Space: the Only Hope for Mankind

Human bodies are in fact living in the three-dimensional universe. But human beings do not understand how the universe goes and where human was originated from. Therefore, they live in the one dimensional world of historical human life. Human rationality and impulse are effected within the limited space. Therefore, the human society is troubled from generations to generations. It could be said that human social life were ruled by a single item of the Tertiary Product: money!

Therefore, recognizing the existence of the three dimensional universe and learning the cosmic truth of human origin are human's only way to saving themselves. Humans' violent actions such as quarrel, murdering, warfare, etc., result from their ignorance of the cosmic origin. People are lured by their immediate interests which incur all sorts of hatred among their neighbors. The hatred can be simply resolved if people expand their vision into the three-dimensional universe and think about their blood lineage connected with the cosmic dust.

In the current world, many people live in big cities, and what they see is basically the cement, automotive, high-rising buildings, and other man-made Tertiary Product. They see fewer and fewer ponds, flowers, trees, birds, and other natural Secondary Product. Thus, the hatred results from their limited vision within the Tertiary Product. The only way to expanding their vision is to let them sit at their offices and read my books.

2.4   Knowledge and Ignorance

Secular humans consider the amount of knowledge on the Tertiary Product as a measure of their wisdom. This makes a lot of innocent people think that they were very ignorant. Friends:

have you ever realized that the knowledge about the Tertiary Product can not be brought from your birth nor be taken with your death. However, your instinct, i.e., your rationality and impulse, is given before your birth. The understanding of the nature is the most needed knowledge in your life. An insight into the truth of the universe is really the wisdom. The people with the wisdom is healthy as well as peaceful. They are calm, serene and natural when facing their death.

The history of human beings on Earth is at least several hundred million years. However, we did not know how human is originated. Human beings appear to be living in a dream. Thanks to the rapid development of natural science over the last centuries, human race can now be waking up. In order to help them wake up from the dream, they need to understand the origin of the universe, the origins of human race and humanity. People should gain more knowledge on the natural science not the Tertiary Product. The knowledge about the nature can never be too much.

If you do not choose to live in the three-dimensional universe then you must live in the one-dimension history of human. If you do not choose to know the origin of the universe then you must return to the dreaming state of human life. In that case, nightmare is not unusual.

2.5   Life and Health

The people who understand the truth of the universe know that their own life is rational and impulsive and derived from the cosmic dust. Earth's ecological system includes other animals, plants, micro-organisms, and other organisms such as bacteria and viruses. They co-exist with the most advanced unique life: human beings. The ecological system is the inevitable result of the rational and impulsive universe, and generated from the cosmic dust.

As the Secondary Product, we should have confidence in our own health because the functioning of human body is rational. The body's immune system is the rational performance of the highest level. However, life is a process of growth from birth to death associated with different levels of impulses which break each stage of equilibrium. These stages also provide the opportunities of abusing, threatening by, for example, harmful viruses. Some threatening opportunities may be fatal. Human body is originated from the dust and is accordingly a vulnerable genetic process. People should gain the knowledge of life, resist most temptations, and follow the way of life. This not only helps people strengthen their own lives but also the lives of future generations.

The universe is very very simple at large-scales (see [14]). The complexity occurs only at small-scales. The unique ecosystem on the Earth's surface is the most complex. Human body can exist only on such surface at small scale. Because the ecosystem may be unique in the universe, we can say that the phenomenon of dreaming occurs only on Earth. By understanding the true three-dimensional universe, human beings will wake up from their mysterious feeling of their life. By then human beings will truly understand and cherish their life and health.

2.6 Marriage and Family

The life of marriage and family is a natural process of natural continuation, which resulted from the rational and impulsive nature of the universe. Therefore, marriage and family should belong to the Secondary Product of the universe. However, today's life of marriage has been impacted by a lot of Tertiary Product such as money. Nearly half of couples go to the ultimate separation. If young lovers recognize the truth of the universe before entering the holy hall of marriage then those cases of impure marriage may happen less commonly. If those people who have entered the stage of marriage are aware of the cosmic truth, their impure marriage may be corrected and purified.

A good life of marriage can help the couples better understand the origin of the universe and human's nature of rationality and impulse. The understanding can help the couples recognize that the roles of husband and wife are equally originated. The male superiority over female is the result of one-dimensional human life. If they know the three-dimensional origin of the universe, the couples can treat each other equally with respect.

2.7 Justice and Causes

The entire universe is just one grand process of justice (rationality) and causes (impulses). Life is a product of justice and causes. Therefore, the progress of human society and the defeat of human crises depend on how social causes serve the justice. For example, the causes of housing, pharmaceutical and medical services must contribute to the maintenance and continuation of human life rather than being driven by the lure of pecuniary interest.

Because they have not understood their origin, human beings have been living in a state of dreaming. Injustice has always existed in human society. To achieve their material interests, a small number of people take advantage of the dreaming mental state of major common people, instead of helping the majority against dreaming.

2.8 Loneliness and Sadness

It is clear that human race is very lonely in the universe and human beings are the passing creature. Thus, human communities should be mutually supportive, rather than mutual persecution. At the same time, we must also understand that human beings are the miracle of the universe originated from the dust. Humans are significantly different from low-level animals. Human beings can develop their own languages and, in turn, understand the universe itself. Human beings can restore their relationship with nature.

If we see the other races of people, we should be happy and friendly with each other rather than with hatred.

A living human body is always the impulsive one. Therefore, people should cherish their own lives, gain the knowledge of the universe, and contribute their passions to the common rational causes.

Chapter 3

Future Humans

3.1 Avoid the Temptation of Mysterious Feeling

It is very natural that young people have infinite curiosity and experience the temptation of mysterious feelings. This kind of curiosity is in fact beneficial. It is the curiosity that leads young people to the knowledge of varied kinds of sciences from kindergarten to graduate school. However, here I talk about the special curiosity which everyone shares. That is the mysterious feeling about the origin of life and the universe. The crises which our society and the natural environment suffer are resulted from human activities driven by the mysterious feeling.

If such visible or measurable materials as galaxies and living life are meaningful (that is, the meaning of rationality and impulse), why do we have to envisage those kinds of unmeasurable ghosts, dark matter and dark energy? Of course, people and scientists did not see the meaning, and accordingly, they need the ghosts, dark matter and dark energy to be fitted into their imagined and false meanings. It is shameless that some people do not believe such kinds of ghosts and dark matter but they promote and disseminate these simply for them to gain personal interests.

The only way adults can live is to learn to protect each other and serve others.

3.2 Future Life is Primarily the Spiritual Life

It is very hard for human authorities to hide the truth of the universe. People will eventually see it. People will eventually get rid of the mysterious feeling on the the origin of life and the universe. Future human life is mainly the spiritual one. We will see the emergence of many new organizations and new careers. These organizations and professionals are "churches" and "preachers". However, their purpose is to promote science, to disseminate the knowledge of microscopic world, chemistry, biology, and the ruling power of the universe (i.e., the rational and impulsive gravity). They help people get rid of the mysterious feelings. They help people live, work, and explore scientifically and rationally. Human race benefits from human social life rather than suffers from the crises brought by human beings.

Appendix A

Can Life Exist without Gravity?

Can life exist without gravity? For example, International Space Station and its on-board astronauts fall freely around the Earth, and suffer no gravity. Can astronauts stay for a long time in the Station?

The original question is provided by some student on Yahoo Answers [16]: "Can life exist without gravity? My physics teacher asked this question today, and we disagreed on the answer. I said no because without gravity stars and stuff wouldn't have formed, and without a star to give us some form of energy to convert. If we don't have energy to convert, then it isn't life. He responded by saying life can exist without gravity as humans and other living things can be sent to space, still live and exist. "

The question is resolved on Yahoo Answers as follows:

"Great Question.

OK- 2 parts (of the answer):

1. On a cosmic scale – You are right.
Without gravity, nothing in our universe would exist at a level higher than coalesced energy, which, even if able to bind at a sub-atomic level, would be unable to bind atomically without the force of gravity which occurs on an atomic scale.

2. On a system specific scale, you are essentially right.
If he was talking about still having the existence of gravity on a universal or cosmic level, and sending life created within a gravitational field into a region without a gravitational field, the following would occur.

The 'life' that we send into a zero gravity region is a system of atoms that was created in and depends on gravity to survive.

NASA's open questions to the public about re-creating earth's gravity in space is an attempt to lengthen the duration of time that a person can stay in space in a man-made optimum environment.

As soon as a person leaves the earth's gravitational field, several things start to occur in the body.

Circulatory problems: within 2-3 days there is a 22% blood volume decrease or loss. Muscle atrophy occurs in space at 5% per week. Bone atrophy occurs in space at 1% per month

Remember, our bodies were developed within the earth's gravity field, our muscles are needed to counter gravity. In space, we no longer need our muscles, so the brain starts to optimize the body by getting rid of what it doesn't need.

Longest stay is 430 days in space.

A human could remain in space probably no more than 2 years without complete irreversible damage.

Our existing atomic system as we know it would change into something else. A person for example, would die. But the atoms and molecules that make up that person don't die, they change their state of existence to create a differently shaped system of atoms. If the system of atoms was provided heat energy, then it would continue to exist and change. Regardless of what system of atoms we started with, a human, a dog, a monkey, a rat, it would not survive as that system. Would it survive as a system of microbials, possibly. How long? 10 yrs, 100 yrs, 1000 yrs, we can't answer that.

Review:

You are essentially correct in both cases. Pending further clarification of his definition of zero gravity, and his definition of life.

Without gravity having existed: no life.

....

Spread the knowledge."

Bibliography

[1] http://web.ipac.caltech.edu/
staff/jarrett/galaxies/spirals.html

[2] http://www.tng.iac.es/
news/2000/07/06/m51/

[3] He, J. & Yang, X. 2009, The Origin Of
Natural Structure (Indiana: AuthorHouse)

[4] He, J. 2003, Astrophys. & Space Sci., **283**,
305

[5] He, J. 2004, Astrophys. & Space Sci., **291**,
163

[6] He, J. & Yang, X. 2006, Astrophys. &
Space Sci., 302, 7

[7] He, J. 2006, Astrophys. & Space Sci., **305**,
197

[8] He, J. 2008, Astrophys. & Space Sci., **313**,
373

[9] He, J. 2005, http://www.arxiv.org/abs/astro-ph/0510535

[10] He, J. 2005, http://www.arxiv.org/abs/astro-ph/0510536

[11] He, J. 2005, http://adsabs.harvard.edu/abs/2005PhDT........17H

[12] Martinez-Valpuesta, I., Knapen, J. H. & R. Buta, R. 2007, Astron. J., **134**, 1863

[13] Eskridge, et al. 2002, Astrophys. J. S., 143, 73

[14] He, J. 2009, The Origin Of Galaxies (Indiana: AuthorHouse)

[15] http://herschel.esac.esa.int

[16] http://answers.yahoo.com/question/index?qid=1006051621689

www.ingramcontent.com/pod-product-compliance
Lightning Source LLC
Chambersburg PA
CBHW050338290526
45785CB00006B/2550